The Geopolitics of Energy into the 21st Century

CENTRAL WYOMING COLLEGE
LIBRARY
RIVERTON, WY 82501

The Geopolitics of Energy into the 21st Century

Volume 1: An Overview and Policy Considerations

A Report of the CSIS Strategic Energy Initiative

Project Cochairs
Sam Nunn
James R. Schlesinger

Project Director
Robert E. Ebel

Project Executive Director
Guy Caruso

Congressional Cochairs
Senator Joseph I. Lieberman
Senator Frank H. Murkowski
Representative Benjamin Gilman
Representative Ellen O. Tauscher

Rapporteur
Elaine L. Morton

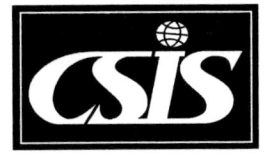

November 2000

About CSIS

The Center for Strategic and International Studies (CSIS), established in 1962, is a private, tax-exempt institution focusing on international public policy issues. Its research is nonpartisan and nonproprietary.

CSIS is dedicated to policy impact. It seeks to inform and shape selected policy decisions in government and the private sector to meet the increasingly complex and difficult global challenges that leaders will confront in this new century. It achieves this mission in four ways: by generating strategic analysis that is anticipatory and interdisciplinary; by convening policymakers and other influential parties to assess key issues; by building structures for policy action; and by developing leaders.

CSIS does not take specific public policy positions. Accordingly, all views, positions, and conclusions expressed in this publication should be understood to be solely those of the authors.

President and Chief Executive Officer: John J. Hamre
Senior Vice President and Director of Studies: Erik R. Peterson
Director of Publications: James R. Dunton

©2000 by the Center for Strategic and International Studies.
All rights reserved.

Library of Congress Cataloging-in-Publication Data

The geopolitics of energy into the 21st century / a report of the CSIS Strategic Energy Initiative ; project cochairs, Sam Nunn, James R. Schlesinger ; project director, Robert E. Ebel.
    p. cm. — (CSIS panel reports)
  Includes bibliographical references.
  Contents: Vol. 1. An overview and policy considerations.—Vol. 2. The supply-demand outlook, 2000–2020.—Vol. 3. The geopolitical outlook, 2000–2020.
    ISBN 0-89206-367-X (set)—ISBN 0-89206-368-8 (v. 1)—ISBN 0-89206-369-6 (v. 2)—ISBN 0-89206-370-X (v. 3)
    1. Energy policy. 2. Energy consumption—Forecasting. 3. Power resources. 4. Geopolitics. 5. International economic relations. I. Nunn, Sam. II. Schlesinger, James R. III. Ebel, Robert E. IV. Center for Strategic and International Studies (Washington, D.C.). V. Series.
HD9502.A2 G465 2000
33.79'01'12—dc21   00-029493

The CSIS Press
Center for Strategic and International Studies
1800 K Street, N.W., Washington, D.C. 20006
Telephone: (202) 887-0200
Fax: (202) 775-3199
E-mail: books@csis.org
Web site: http://www.csis.org/

Contents

List of Tables and Figures	vi
Acknowledgments	vii
Participants	ix
Abbreviations	xiii
Executive Summary	xv
1. Introduction	1
2. Energy Outlook	4
3. Geopolitics and Energy: A Symbiotic Relationship	7
4. Energy and Geopolitics: Policy Considerations	19
5. Conclusions	30

List of Tables and Figures

Figure 2.1. Predicted Changes in World Consumption of Energy, 2000 and 2020   5

Figure 3.1. Predicted Changes in World Generation of Electricity, 2000 and 2020   12

Figure 3.2. Growth of Energy Consumption, 2000–2020   17

Table 4.1. Predicted Changes in Oil Production Capacity, 2000–2020   23

Acknowledgments

The CSIS Strategic Energy Initiative (SEI) was launched in July 1998 in response to the concerns of former senator Sam Nunn and former secretary of energy James R. Schlesinger, who believed that a lengthy period of secure energy supplies had led U.S. policymakers to neglect the changing relationship between geopolitics and energy at the turn of the century. This changing relationship will require a rethinking of U.S. foreign policies, environmental policies, and the broader national security strategy. The three-volume study, *The Geopolitics of Energy into the 21st Century*,[*] responds to these concerns.

Experts from the private and public sectors and from academe were brought together to discuss and analyze future energy supply and demand and to balance this analysis with a review of the key oil- and gas-exporting countries of the world. Guy Caruso took on the difficult task of day-to-day management of SEI, ably supported by Megan Smith in her position as project coordinator. Elaine O. Morton as rapporteur had perhaps the most difficult assignment, that of weaving together what is a complex story. She succeeded in every regard and is recognized as the principal author of all three volumes.

The project benefited immensely from the continued support and involvement of Anthony A. Smith, then CSIS executive vice president and chief operating officer, and Robin Niblett, then vice president for strategic planning. Their encouragement was essential to SEI's successful conclusion. Equal thanks are due the many CSIS program directors and other CSIS staff for their contributions to volume 3, *The Geopolitical Outlook:* Gerrit Gong, Teresita Schaffer, Simon Serfaty, Lowell Fleischer, Howard Wiarda, Anthony Cordesman, Judith Kipper, Keith Bush, Connie Freeman, and Steve Morrison.

A number of individuals offered their insight and judgments throughout the project. We particularly recognize Bill Hale, Arnold Baker, Wilfrid Kohl, Herman Franssen, and Jim Ragland. To each of the task force members, without whose enthusiastic participation SEI would not have been possible, and to Lisa Hyland, who took over upon the departure of Megan Smith, we extend our personal thanks.

Special recognition must be given to congressional cochairs Senators Frank H. Murkowski (R-Alaska) and Joseph I. Lieberman (D-Conn.) and Representatives Benjamin Gilman (R-N.Y.) and Ellen O. Tauscher (D-Calif.). Their continued support, and that of their staff, was essential to a successful SEI.

---

[*] The three-volume study includes *An Overview and Policy Considerations* (volume 1), *The Supply-Demand Outlook, 2000–2020* (volume 2), and *The Geopolitical Outlook, 2000–2020* (volume 3). Although the SEI Advisory Committee has reviewed volume 1, responsibility for the contents rests with CSIS, as it does for the two companion volumes.

Finally, the remarkable leadership of cochairs Sam Nunn and James R. Schlesinger throughout the project must be noted. Their personal involvement and guidance have been of tremendous importance.

CSIS is deeply appreciative of the efforts of all participants to produce a report that is timely, objective, and bipartisan and that carries policy considerations serving the best interests of the United States and of energy producers and consumers everywhere.

The geopolitics of energy is rarely static. Events of the day carry implications for energy supply, consumption, and prices—sometimes immediate, sometimes delayed, sometimes hidden. By attempting to define, in advance, the form that these events might take—and the resulting impact on energy—this report may remove some surprise from the future and ease the way for decisionmakers in both the public and private sectors.

Robert E. Ebel
Director, Energy and National Security
CSIS

Participants

Project Cochairs

Sam Nunn
James R. Schlesinger

Congressional Cochairs

Senator Joseph I. Lieberman (D-Conn.)
Senator Frank H. Murkowski (R-Alaska)
Representative Benjamin Gilman (R-N.Y.)
Representative Ellen O. Tauscher (D-Calif.)

Advisory Committee

Rodney F. Chase
Deputy Group Chief Executive
BP

Charles Fabrikant
Chairman, President & CEO
SEACOR SMIT, Inc.

Luis Giusti
Former Chairman and CEO
Petroleos de Venezuela, S.A.

Richard K. Gordon
Vice Chairman, Investment Banking
Merrill Lynch

Robert L. Healy
Vice President, Federal Government Relations
ARCO

Dale P. Jones
Former Vice Chairman
Halliburton Company

Diana MacArthur
Chair and CEO
Dynamac Corporation

Jessica T. Mathews
President
Carnegie Endowment for International Peace

John V. Mitchell
Chairman, Energy and Environment Program
The Royal Institute of International Affairs

David M. Nemtzow
President
The Alliance to Save Energy

Johan Nic Vold
Executive Vice President
Statoil

C. Paul Robinson
President
Sandia National Laboratories

William Ruckelshaus
Former Chairman
Browning-Ferris Industries

John W. Snow
Chairman, President & CEO
CSX Corporation

Robert E. Wilhelm
Director and Senior Vice President
(retired)
Exxon Mobil Corporation

Daniel B. Yergin
President
Cambridge Energy Research Associates

Program Director

Robert E. Ebel

Project Executive Director

Guy Caruso

Rapporteur

Elaine L. Morton

Project Coordinator

Megan Smith

Task Force Members

A total of eight task forces were assembled, from experts in both the private and public sectors, to carry out the work of the Strategic Energy Initiative project. Task forces were established on Asia, the Middle East, the former Soviet Union, the Western Hemisphere, Africa, Europe, the Environment, and Technology. Although many of the experts served on several task forces, their names are presented here in a combined list so as to avoid redundancy. In addition, those participants who offered their insights on a broader basis and were not members of any particular task force are also recognized.

Art Andersen
U.S. Department of Energy

Kai Anderson
Office of Senator Joseph Lieberman

Arnold Baker
Sandia National Laboratories

David D. Bosch
Aramco Services Company

John Brodman
U.S. Department of Energy

Jim Bruce
Wiley, Rein, & Fielding

Todd Burger
Office of Representative Benjamin Gilman

Ellison Burton
Dynamac Corporation

G. Daniel Butler
U. S. Department of Energy

Alys Campaigne
Office of Senator Joseph Lieberman

Michael Canes
American Petroleum Institute

Galen Cobb
Halliburton Company

Len Coburn
U.S. Department of Energy

Melissa Coffey
Office of Representative Benjamin Gilman

Barry Cohen
Energy Information Agency

Bud Coote
Central Intelligence Agency

Terry Coyne
U.S. Government

Peter A. Davies
BP

Ged Davis
Shell International Limited

Fred Downey
Office of Senator Joseph Lieberman

Luther W. Dudich
Alliance to Save Energy

Charles Ebinger
International Resource Group

Lana Ekimoff
U.S. Department of Energy

Marvin Fertel
Nuclear Energy Institute

George Folsom
Koch Industries

Herman Franssen
International Energy Associates

Chas Freeman
Middle East Policy Council

Steve Gallogly
U.S. Department of State

Larry Goldstein
Petroleum Industry Research Foundation

Robert L. Gould
CSX Corporation

William Hale
Exxon Mobil Corporation

Dan Hickey
Texaco

Lynn Hogan
Arco

Karen Hunsicker
Senate Committee on Energy and Natural Resources

Maureen Koetz
Nuclear Energy Institute

Wil Kohl
*SAIS
Johns Hopkins University*

Michael V. Kostiw
Texaco

Peter Lyons
Office of Senator Pete V. Domenici

Neil McKeown
U.S. Government

Charles McPherson
The World Bank

Eric Melby
Scowcroft Group

Jack Moore
Aramco Services Company

Angela K. Morin
Alliance to Save Energy

Ed Morse
Hess Trading Company

Steven Mullen
U.S. Department of State

Peter Muller
Office of Representative Ellen Tauscher

Shirley Neff
Senate Committee on Energy and Natural Resources

Deanna Okun
Office of Senator Frank Murkowski

Willy H. Olsen
Statoil

Michelle R. Perez
Alliance to Save Energy

James Placke
Cambridge Energy Research Associates

Jerry Pruzan
Arco

James Ragland
Aramco Services Company

Joyce Rechtschaffen
Office of Senator Joseph Lieberman

Richard Sawaya
Arco

Robert Shinn
CSX Corporation

Adam L. Shrier
International Consulting

Adam E. Sieminski
BT Alex Brown

Ron Simard
Nuclear Energy Institute

Elizabeth A. Sollie
Texaco

Vito Stagliano
Commonwealth Edison Company

Cyrus Tahmassebi
Energy Trends, Inc

Michael Townshend
BP

Geir Westgaard
Statoil

Guenther O. Wilhelm
Exxon Corporation (retired)

Abbreviations

The following abbreviations are used in the three-volume study.

Bbbl	billion barrels
bbl	barrels
Bcf/d	billion cubic feet per day
Bcm	billion cubic meters
BkWh	billion kilowatt hours
Bst	billion short tons
Bt	billion metric tons
Btu	British thermal unit
kW	kilowatt
kWh	kilowatt hour
MMbbl/d	million barrels per day
MMst	million short tons
MMt	million metric tons
mpg	miles per gallon
MW	megawatt
quad	quadrillion Btu
st	short ton
t	metric ton
Tbbl	trillion barrels
Tcf	trillion cubic feet
Tcm	trillion cubic meters

Numbers used in this report are defined according to the system used in the United States.

million	10^6 (6 zeros)
billion	10^9 (9 zeros)
trillion	10^{12} (12 zeros)
quadrillion	10^{15} (15 zeros)

Executive Summary

The Center for Strategic and International Studies (CSIS) launched its Strategic Energy Initiative (SEI) in mid-1998 on the premise that the benign global energy situation that had prevailed since the late 1980s masked two dangers.

First, it obscured significant geopolitical shifts both ongoing and forthcoming that could affect future global energy security, supply, and demand.

Second, it led to complacency among policymakers and the public about the need to incorporate long-term global energy concerns into near-term foreign policy decisions.

By midyear 2000 the state of the world oil market had undergone considerable turbulence, marked by rapidly rising oil prices as oil-exporting countries were benefiting from staged reductions in production that had been initiated more than two years earlier. The delicate balance between supply and demand was demonstrated once again.

Instead of dwelling on the oil market turbulence in 2000, however, this report assesses the international energy supply-and-demand relationships likely to prevail in the first two decades of the twenty-first century, highlighting the different ways that geopolitical developments could affect global energy markets between 2000 and 2020. In light of the world's future energy needs, this report series also points out the contradictions inherent in certain of the energy objectives and foreign policies pursued by the United States and other Western governments. Finally, the report offers policy considerations that, if implemented, could help ensure that energy supplies are adequate to meet projected worldwide demand, are not excessively vulnerable to major interruptions, and are produced in ways that minimize damage to the environment.

It may appear that parts of this assessment are unduly pessimistic, that positive factors have been overlooked. These SEI assessments do stress prospects for instability and for interference in energy supplies, but only to alert policymakers about the fragility of reliable and timely supplies.

Energy Outlook to 2020

During the next 20 years, providing there is no extended global economic dislocation, energy demand is projected to expand more than 50 percent. This growth will be unevenly distributed, with demand increasing in the industrialized world by some 23 percent while more than doubling, from a much lower base, in the developing world, with Asia accounting for the bulk of this increase. At some point during this period, the developing world will begin to consume more energy than the developed world. Energy supply will need to be expanded substantially to meet this demand growth. Although the Persian Gulf will remain the key marginal oil supplier, all producing countries must contribute to supply to the extent they can.

Central to the geopolitics of energy during 2000–2020 is the fact that energy demand will be met in essentially the same ways as it was met at the end of the twentieth century. Fossil fuels will provide the bulk of global energy consumption, rising marginally from an 86 percent share in 2000 to an 88 percent share in 2020. Although oil will dominate global energy use and coal will retain its central role in electricity generation, natural gas use will increase noticeably. Indeed, the relative contributions of oil and coal to world energy consumption will actually decline whereas only natural gas will demonstrate a growth in both absolute and relative terms. Nuclear power will decline in both relative and absolute terms; renewables, including hydropower, and alternative energy sources, while growing in absolute terms, will not capture a greater relative share of the market.

Development of oil and gas reserves is judged sufficient to meet projected global demand well beyond this period. The most noticeable trend during 2000–2020 will be the growing mutual dependencies between energy suppliers and consumers. Key aspects of this trend, which are set out below, may appear rather obvious—and they are; how to respond in today's changing environment is much less so.

- The Persian Gulf will remain the key marginal supplier of oil to the world market, with Saudi Arabia in the unchallenged lead. Indeed, if estimates of future demand are reasonably correct, the Persian Gulf must expand oil production by almost 80 percent during 2000–2020, achievable perhaps if foreign investment is allowed to participate and if Iran and Iraq are free of sanctions.

- While the Persian Gulf's share of world oil production continues to expand, the share of North America and Europe, the world's most stable regions, is projected to decline.

- The share of world oil production from the former Soviet Union is projected to increase from 9 percent to almost 12 percent. But, as had been the case in earlier years, this oil will follow the market, not attempt to lead it.

- The Caspian oil contribution to world supply will be important at the margin but not pivotal.

- Asian dependence on Persian Gulf oil will rise significantly, and the resulting necessity for longer tanker journeys will put more oil at risk in the international sea lanes.

- European dependence on Persian Gulf oil will remain significant.

- The European need for natural gas will be covered by a handful of suppliers, Russia being the most significant, which underscores a worrisome dependency.

- U.S. net oil imports will continue their steady growth.

- Anticipated growth in the use of natural gas—in considerable part engendered as a fuel for electric power stations—raises a new series of geopolitical issues, leading to new political alignments.

- Electricity will continue to be the most rapidly growing sector of energy demand; developing economies in Asia and in Central and South America will show the greatest increase in consumption. The choice of primary fuel used to supply power plants will have important effects on the environment.

- Technological change and improvements in energy efficiency have made their mark on recent energy supply-and-demand balances. Future energy supply and demand must reflect not only a continuation of these successes but an acceleration wherever possible.

Geopolitics and Energy: A Symbiotic Relationship

How Might Geopolitics Affect Energy?

Four main geopolitical trends are likely to influence energy supply and demand during the years ahead.

THE CONTINUING DOMESTIC FRAGILITY OF KEY ENERGY-PRODUCING STATES. The world drew some portion of its energy supplies from unstable countries and regions throughout much of the twentieth century. By 2020, fully 50 percent of estimated total global oil demand will be met from countries that pose a high risk of internal instability. A crisis in one or more of the world's key energy-producing countries is highly likely at some point during 2000–2020.

GLOBALIZATION. Economic globalization will impose new competitive and political pressures on many of the world's leading energy producers and consumers. It will serve as a spur for growth in global energy supply and demand. It could also lead to serious swings in energy prices and demand because country-specific or regional recessions or other influencing events can now be transmitted quickly around the world. In such a globalized world, energy producers and consumers will become ever more sensitive to their mutual interdependence.

THE GROWING IMPACT OF NONSTATE ACTORS. This impact will be evident in three distinct areas. First, adroitly employing new information technologies, nongovernmental organizations (NGOs) will play a growing role in defining the ways that energy is produced and consumed. Second, terrorist groups, with access to the same technologies, will be in a position to inflict great operational damage on increasingly complex energy infrastructures. Third, radical activists will be in a position to disrupt operational infrastructure through cyberterrorism.

CONFLICT AND POWER POLITICS. The potential for armed conflict in energy-producing regions will remain high. Early in the twenty-first century, as a result, a weakening of U.S. alliance relationships in Europe, the Persian Gulf, or Asia could have major impacts on global energy security. U.S. concerns over the proliferation of weapons of mass destruction (WMD) and the desire to promote democratization and market liberalization around the world will also have a significant effect on key energy exporters. The future viability of the energy-producing states in the Caspian and Central Asia will be shaped by the competing objectives or interests of Russia, the United States, and adjacent regional powers.

How Might Energy Affect Geopolitics?

There are five main ways in which energy may affect geopolitical outcomes:

SWINGS IN ENERGY DEMAND. A dramatic decline in global energy consumption, brought on by economic recession, could trigger instability in many of the world's major energy-exporting countries. Conversely, continued economic growth, accompanied by rising energy demand, would place more power in the hands of the exporters.

SWINGS IN ENERGY SUPPLY. Just as demand is vulnerable to sharp shifts up or down, so is supply. If discovery and development of new reserves and the addition of producing capacities match demand growth, an acceptable balance between supply and demand can be maintained. But a number of factors must be satisfied if supply growth is to be encouraged, including an attractive host-country investment climate and the opportunity for acceptable investment returns. At the same time, political events and logistical interruptions can interfere with supply.

COMPETITION FOR ENERGY IN ASIA. As countries in Asia seek to secure growing levels of energy imports, two geopolitical risks emerge. First, historical enmities might boil over into armed conflict for control of specific energy reserves in the region. Second, the rising dependence of China on Persian Gulf oil could well alter political relationships within and outside the region. For example, China might seek to build military ties with energy exporters in the Persian Gulf in ways that would be of concern to the United States and its allies.

ENERGY AND REGIONAL INTEGRATION. Energy infrastructure projects may serve to strengthen bilateral economic and political ties in certain instances. In Asia, for example, energy networks, along with trade liberalization, could serve to reduce historical tensions and place Asian economic growth on a firmer footing. Similar forces might come into play in Europe, linking Russia to the European Union (EU); in South Asia, drawing Bangladesh and India closer together; and in the Far East, linking Russia and China.

ENERGY AND THE ENVIRONMENT. Environmental concerns will have an increasingly important geopolitical bearing on energy decisionmaking by governments, by producers, and by consumers in the next decades. Should governments pursue aggressive strategies for reducing carbon emissions, a new political fault line could emerge between developed and developing countries.

Policy Contradictions and Considerations

The interplay of geopolitics and energy early in the twenty-first century is at the root of an array of complex policy challenges that governments around the world must now confront. The three interlocking policy challenges are to ensure that (1) in the long term, supplies will be adequate to meet the world's energy needs; (2) in the short term, those supplies are reliable and not subject to serious interruptions; and (3) at all times, energy is produced and consumed in environmentally acceptable ways.

Energy Availability

U.S. policy today contains a fundamental contradiction. Oil and gas exports from Iran, Iraq, and Libya—three nations that have had sanctions imposed by the United States or international organizations—are expected to play an increasingly important role in meeting growing global demand, especially to avoid increasing competition for energy with and within Asia. Where the United States imposes unilateral sanctions (Iran and Libya), investments will take place without U.S. participation. Iraq, subjected to multilateral sanctions, may be constrained from building in a timely way the infrastructure necessary to meet the upward curve in energy demand. If global oil demand estimated for 2020 is reasonably correct and is to be satisfied, these three exporters should by then be producing at their full potential if other supplies have not been developed.

History has demonstrated that unilateral sanctions seldom are successful in persuading nations to alter their behavior. Multilateral sanctions provide a broader front and a greater guarantee of success. Multilateral sanctions test the ability and willingness of enforcing nations to hold together for the duration, however, while both multilateral and unilateral sanctions are viewed as targets of opportunity for the entrepreneurial trader.

Western governments should avoid the indiscriminate use of sanctions. The value of multilateral sanctions should be weighed against the value of engagement and dialogue. When the use of sanctions is deemed admissible in the support of international interests, governments should adopt a graduated approach and make every effort to ensure that the coverage of the sanctions is as targeted as possible. This should include a cost–benefit analysis of whether curtailing investment in, or revenue from, energy production will genuinely dissuade the target government from the specific behavior that provoked the imposition of sanctions.

Despite a limited success record, sanctions will continue to be used as a tool of foreign policy—as a means of rejecting the conduct of a particular nation—simply because there are no acceptable alternative courses of action. The world will have to live with the inherent limitations of the sanctions.

> **Policy consideration:** Avoid the indiscriminate use of sanctions. The value of multilateral sanctions should be weighed against the value of engagement and dialogue. When the use of sanctions is deemed admissible in the support of international interests, ensure that the coverage of sanctions is as targeted as possible. Unilateral sanctions are not an effective policy tool.

A similar contradiction exists in U.S. policy toward the Caspian region and Central Asia, where the United States is committed to reinforcing the newly independent states but where contrasting U.S. policies toward Iran, Turkey, and Russia are likely to influence, rightly or wrongly, the construction of commercially viable pipelines for the export of Caspian oil and gas. A policy approach that ties exports primarily to one pipeline route—with the goal of avoiding Iran and Russia as transit states—before the political and economic viability of that route is known may undercut the pace of energy development in the region, to the dismay of both producing states and potential transit states.

Oil and gas exports from the Caspian region and Central Asia hold the prospect of becoming a valuable additional source of energy supply. Even as the U.S. government works to make feasible an East–West transportation corridor that bypasses Russia and Iran, the United States should not obstruct the development of alternative routes that would ultimately offer exporters a diverse and economically attractive set of options for transporting oil and gas to foreign markets, especially those markets in Asia and the Far East.

> **Policy consideration:** Do not obstruct the development of economic routes that would ultimately offer Caspian and Central Asian exporters a diverse set of options for transporting oil and gas to foreign markets.

Beyond these contradictions, if Western governments are to ensure adequacy of supply early in the twenty-first century, policies must be framed toward encouraging energy-producing countries to open their energy sectors to greater foreign investment. This would include provisions for the enforcement of contracts, guarantees for private property, anticorruption measures, and stable fiscal regimes. Increased private investment must occur as early as possible in exploration and production facilities and in transportation infrastructure, especially in Asia, if the world's energy supplies are to reach markets in sufficient quantities during the 2010–2020 period.

> **Policy consideration:** Encourage energy-producing countries to ensure that their energy sectors attract and support greater foreign investment.

Given the continuing importance of a small group of energy-producing and -exporting countries to the future health of the global economy, it is vital that the United States and other Western governments place diplomatic relations, trade policies, and foreign assistance programs with each of these countries at or near the top of policy priorities.

It is in the self-interest of the United States and other Western governments to support China—rapidly emerging as a major oil importer—as it diversifies its sources of and forms of imported energy and encourage China to not rely excessively on the Persian Gulf. China is considering development of an infrastructure to support oil and gas imports from Russia and Central Asia and also for transit onward to other countries in the Far East. Collaborative cross-national energy infrastructure projects can play an important role in lessening the risks of future conflict over energy resources. However, such energy linkages may not always be in the best political interests of the United States.

Energy Reliability

In the early decades of the twenty-first century, because burgeoning energy demand must be met largely by a small number of oil and gas suppliers and because supply routes are lengthening, the risk posed by supply interruptions will be greater than it was at the end of the twentieth century.

Military conflict will remain a threat to most energy-producing regions, particularly in the Middle East where almost two-thirds of the world's oil resources are located. In addition, domestic turmoil within the key energy-producing countries constitutes another threat to reliability of energy supplies. At least 10 of the 14 top oil-exporting countries run the risk of domestic instability in the near to middle term.

The United States should retain as far as possible its ability to defend open access to energy supplies and international sea lanes. At a time when the administration faces myriad competing demands for military and peacekeeping interventions, this mission should be considered a strategic priority and may call for greater emphasis on, and increased investment in, appropriate military capabilities.

> **Policy consideration:** The United States should retain as far as possible its ability to defend open access to energy supplies and international sea lanes.

Some observers are concerned that the United States may seek relief from its self-imposed responsibility as the protector of the world's sea lanes, which are used for the transport of fuels and are becoming more crowded. U.S. allies in Europe and Asia should be prepared to shoulder a greater share of the financial cost of protecting energy supply, including sea-lane protection.

> **Policy consideration:** U.S. allies in Europe and Asia should be prepared to shoulder a greater share of the financial cost of protecting energy supply, including sea-lane protection.

No protector comparable with the U.S. role on the high seas exists for the increasingly important long-distance pipeline infrastructure. At a government-to-government level, international agreements to protect pipeline systems might have a deterrent effect. Governments must also find ways to work with the private sector to minimize the vulnerability of all energy infrastructures to sabotage or terrorist attack. Cyberterrorism may well pose the greatest threat during the time period under review.

> **Policy consideration:** Governments must find ways to work with the private sector to minimize the vulnerability of energy infrastructure to sabotage or terrorist attack, including cyberterrorism.

The more feasible approach in the near to medium term to mitigate the risks of gas-supply interruptions is to encourage importing countries to promote diversity among suppliers and delivery routes. European governments, particularly in view of their high dependence on Russian gas, should look closely at how security of gas supply might be enhanced.

To meet these challenges to reliable supply, importing nations must engage in contingency planning. The practice of holding government-financed strategic petroleum reserves is one essential method of limiting the impact of supply interruptions, provided that the stocks held are truly reserved for the intended purpose and not for manipulating domestic prices. Governments should maintain and, where appropriate, expand government-financed and -controlled strategic petroleum reserves. This could include extending the International Energy Agency (IEA) emergency preparedness program to nonmember countries that will become major oil importers and supporting the concept of regional stabilizing initiatives. For the foreseeable future, however, it would appear to be impractical and prohibitively expensive to hold strategic natural gas reserves.

> **Policy consideration:** Governments should maintain and, where appropriate, expand government-financed and -controlled strategic petroleum reserves, reserving their use for supply interruptions.

Energy and the Environment

Energy production and use have become linked to environmental concerns. Air pollution, oil spills, and their impact on habitats are among the many challenges confronting government and the energy industry.

However, the energy industry's primary source of international friction may revolve around the issue of global climate change, as amply demonstrated by the contentious debate over the cost and benefits of the Kyoto Protocol.

The United States is unlikely to ratify the Kyoto Protocol in its present form. Clearly, global climate change can potentially have major implications for the economies of the world. Continued research and understanding of the facts are imperative for progress on this issue.

By 2020, energy consumption by the developing countries of the world is expected to exceed energy consumption by the developed countries. This may hold particular implications for the environment. Technologies must be made available to help ensure that, for developing countries, the burning of fossil fuels releases minimal pollutants. Moreover, fuel choices must be broadened to include cost-competitive nuclear electric power.

There will be no easy solutions. Clean-coal technology stands beyond the economic reach of most developing countries. Switching from coal to natural gas will take time inasmuch as deliveries will be dependent on the availability of costly long-distance natural gas pipelines and liquefaction and regasification facilities for the export and import of liquefied natural gas.

> **Policy consideration:** Economically and environmentally sound technologies must be made available to help developing countries meet increasing energy demands.

Nuclear power is emissions free but poses its own set of competing policy concerns, ranging from reactor safety to waste disposal and nuclear weapons proliferation. Western governments should assess the conditions under which nuclear power could make a significant contribution to electricity supply in the developing world by first assessing those conditions under which nuclear power could make a continuing contribution to their own supply.

Developing country decisionmakers would have to ask themselves, "Is this the most sensible answer to our power problems, and is this option reasonably affordable?" Three essential criteria for a fourth-generation nuclear power reactor, suitable above all for use in developing countries, would have to be met.

- Modular construction, with a generating capacity of approximately 100 MW;

- Cost competitive compared with fossil-fuel generating plants; and
- Proliferation resistant.

   Policy consideration: Western nations should assess the conditions under which nuclear power could make a significant contribution to electricity generation in the developing world.

A major challenge for the future is quite evident: how to produce, transport, and burn fossil fuels in massive amounts but in an environmentally friendly manner. Is that possible only through technological breakthrough? Because in democratic countries the regulation and deregulation process can involve lengthy legislative and executive interaction and a complex public vetting process, simply recommending that policymakers eliminate those regulations that inhibit bringing technological innovation to market is meaningless. Instead, Organization for Economic Cooperation and Development (OECD) governments should expand basic research leading to more efficient fuel use and to viable alternative fuels. At the same time, governments should fashion regulatory processes and standards that favor the market success of environmentally friendly innovative energy technology.

Countries should review the extent to which subsidies for domestic energy sectors are inconsistent with their global energy policies.

   Policy consideration: OECD governments should expand basic research on energy technologies; concurrently, policymakers should eliminate those environmental regulations that inhibit bringing technological innovation to market. All governments should review the extent to which domestic energy subsidies are inconsistent with global energy policies.

Three Broad Conclusions

Three broad conclusions can be drawn from this analysis of geopolitics of energy into the twenty-first century.

- The United States, as the world's only superpower, must accept its special responsibilities for preserving worldwide energy supply.

- Developing an adequate and reliable energy supply to realize the promise of a globalized twenty-first century will require significant investments, and they must be made immediately.

- Decisionmakers face the special challenge of balancing the objectives of economic growth with concerns about the environment. This challenge has multiple parts: finding ways to increase security and reliability of supply; ensuring greater transparency in energy commerce; and strengthening the role of international institutions in matters of energy and the environment.

One of the ironies at the turn of the century is that, in an age when the pace of technological change is almost overwhelming, the world will remain dependent, during 2000–2020 at least, essentially on the same sources of energy—fossil fuels—

that prevailed in the twentieth century. Political risks attendant to energy availability are not expected to abate, and the challenge for policymakers is how to manage these risks.

What's New?

The influence of nongovernmental organizations (NGOs) on public and private energy-related policy decisions is perceived to be expanding.

Projected energy consumption in developing countries will begin to exceed that of developed countries, a change that will carry political, economic, and environmental considerations.

The spread of information technology and use of the Internet dramatically change the way business is conducted, and this change carries with it a new set of vulnerabilities.

The prospects of cyberterrorist attacks on energy infrastructure are very real; such attacks may be the greatest threat to supply during the years under review.

Global warming is attracting growing attention, and that attention will likely shape debate on future energy policies; it is hoped that debate will reflect sound science and factual analysis.

Security of Supply

If U.S. military power is committed to a limited but extended protection effort in Northeast Asia, the capacity to respond to a crisis like that of 1990 in the Persian Gulf will be severely limited. The United States will need to rebalance its security relations.

Policy Contradictions

The greater need for oil in the future is at odds with current sanctions on oil exporters Libya, Iraq, and Iran.

The United States deals with energy policy in domestic terms, not international terms; U.S. energy policy is therefore at odds with globalization.

CHAPTER 1

Introduction

At the beginning of the twenty-first century, much of the world seemed poised to enjoy a period of unprecedented prosperity. The end of the Cold War had ushered in an era when no major war seems imminent. Globalization had unlocked economic growth in the industrialized world as well as in many developing nations. Underpinning this increasingly confident perspective was the fact that the world had become accustomed to plentiful and reliable energy supplies. The supply interruptions of 1973 and 1979, bringing with them "car-less" Sundays in Europe and long lines at the gas pumps in the United States, were distant memories. National security concerns about oil had all but disappeared from public view.

Then, beginning in 1997, oil supply substantially exceeded demand and triggered a collapse in world oil prices, which ultimately fell to a low of less than $10 per barrel. Member countries of the Organization of Oil Exporting Countries (OPEC) and other oil exporters sought relief by working toward a common goal: to raise prices by taking oil off the world market. This willingness to work together to reduce supplies came from a coincidence of interests. National budgets had been badly hurt by the oil price declines of 1998–1999. Adherence to agreed production cuts was surprisingly successful, to the extent that between March 1999 and September 2000 prices had more than tripled, exceeding $37 per barrel, despite incremental restorations of the supply during 2000 by oil exporters in an attempt to bring prices back to more acceptable levels. In the judgment of some, an energy crisis had now descended upon consumers.

Before the price collapse and resurgence, however, the passage of time, growing global prosperity, and a comparatively benign world energy situation had produced an atmosphere of relative calm—a striking contrast with the perception of crisis that was widespread during the 1970s and early 1980s. This sense of calm masked two dangers:

- It obscured potentially significant underlying geopolitical shifts that are likely to affect future global energy security, supply, and demand.

- It led to complacency on the part of policymakers and the public about the need to incorporate long-term energy concerns into near-term foreign policy decisions.

Whether the promise of globalization is realized will depend in large measure on whether energy can be supplied in amounts sufficient to sustain a high rate of economic growth. More specifically, energy supplies must be adequate to meet projected worldwide demand, priced at mutually acceptable levels, delivered to consumers in an uninterrupted manner, and used in ways that will minimize damage to the environment.

Whether these desirable energy goals will be achieved during 2000–2020 will be influenced greatly by geopolitics. The lessons are there: the Organization of Arab Petroleum Exporting Countries (OAPEC) oil embargo in response to U.S. support of Israel during the 1973 October War; the sharp increases in oil prices that occurred during the Iranian revolution in 1979, during the initial phase of the Iran-Iraq War in 1980, and immediately following Iraq's invasion of Kuwait in 1990. Clearly, a major conflagration in the Middle East—or indeed serious political instability within a major oil-exporting nation—would almost certainly lead to dangerous oil supply interruptions with accompanying price escalation and global economic damage.

Although the potential impact of geopolitical events on energy supply and reliability is not difficult to envision, it is equally true that energy issues have often exercised a reciprocal influence on geopolitical outcomes. The 1991 Gulf War provides a vivid example both in Iraq's decision to invade Kuwait, precipitated in large measure by their dispute over oil prices, and in the Western response in which oil security was a powerful incentive. More recently, U.S. policymakers have hoped that rapid and effective development of energy resources in the Caspian region and Central Asia will enable the former Soviet republics in the region to solidify economic and political independence from Russia.

It is important to recognize, however, that the same interplay of geopolitics and energy that shaped energy security, supply, and demand in the latter part of the twentieth century may not prevail in this century. First, the world will need to adapt to a radically altered security context, with the United States finding itself the sole superpower but often reluctant to accept the responsibilities and consequences of this role. Second, spurred by globalization, rapid economic growth in the developing world will drive world energy demand upward, particularly in Asia, thus creating new regional and geopolitical tensions. Although the global economy will have little choice but to rely primarily on fossil fuels to meet burgeoning demand, environmental concerns, especially regarding global warming, will influence corporate and governmental actions in the future far more than they have in the past. Nongovernmental organizations (NGOs), taking advantage of the new information technologies, will play an increasingly important role in conveying environmental and other concerns to the public and to governments alike.

The interplay of geopolitics and energy is at the root of an array of difficult policy choices that nations must now confront. Competing national goals already are coming into conflict as policymakers seek to elaborate coherent and consistent foreign, national security, environmental, and energy policies. For example, U.S. foreign policy objectives that have led to sanctions may be in conflict with the need to ensure that adequate energy supplies are available in 2000–2020. The controversy surrounding the Kyoto Protocol is an obvious example of the distinctly different environmental, energy, and economic priorities of developed and developing countries.

With these factors in mind, CSIS in July 1998 launched the SEI, a comprehensive study designed to examine the relation between energy and geopolitics—as it exists at the beginning of the twenty-first century and as it is likely to unfold during the first two decades of the century. The SEI brought together government officials, industry executives, and academic and policy experts over an 18-month period to discuss and analyze the likely interrelationships between energy and geopolitics over the first two decades of the twenty-first century. Specifically, SEI set out to

- Review energy supply-and-demand relationships, as they are projected to develop during the first two decades of the twenty-first century,

- Analyze geopolitical opportunities and constraints as they relate to energy, and

- Sharpen understanding of the policy contradictions inherent in attempting to achieve sometimes incompatible goals in energy, foreign affairs, national security, and environmental policy.

CHAPTER 2

Energy Outlook

During the period from 2000 to 2020,[1] worldwide energy demand is expected to increase by more than 50 percent, as illustrated in figure 2.1. This projected increase in energy demand will be unequally distributed among countries at different stages of economic development. In the industrialized countries, energy consumption will rise by only 23 percent, but in the developing countries, energy consumption, while proceeding from a much smaller base, will more than double. Fossil fuels (coal, natural gas, and crude oil) will continue to be the primary forms of energy used to meet worldwide demand—rising from an 86 percent share in 2000 to an 88 percent share by 2020. The role of nuclear power, constrained by both negative public perceptions and safety and capital cost considerations, is projected to decline and will cover only 4 percent of total energy demand by 2020. Hydropower generation will remain limited by the exhaustion of appropriate sites for hydroelectric dams and by growing environmental objections to such large water resource projects. Although there will be growth in absolute terms in other renewable sources of energy—solar, biomass, wind—as well as alternative energy sources (for example, fuel cells), their relative share of energy consumed is not expected to change measurably in the time frame under consideration.

Among the three major fossil fuels, crude oil will continue to predominate and will account for a 36 percent share of energy consumed in 2020. Natural gas will follow at 29 percent, and coal will account for 23 percent. Each of these three fossil fuels will increase in absolute levels of production. Oil use will rise significantly in support of the expanding motor vehicle sector. Natural gas use is expected to rise even faster than oil, particularly in the developing countries and especially in recognition of its value as a clean and cost-attractive fuel for electricity generation. Coal, benefiting from a large resource base and entrenched political support, will remain dominant in electric power generation.

There will be increasing reliance on market mechanisms and the private sector to find resources, develop new technologies, and invest in new industrial capacity. Market forces (including, but not limited to, deregulation and restructuring in grid-based energy industries such as electric power and natural gas), significant technological developments in methods used for the exploration and production of oil and natural gas, and better information support the conclusion that supplies of crude oil, natural gas, and coal will be sufficient to cover demand at least through

---

1. For the detailed analysis upon which these conclusions are based, see the companion SEI report, volume 2 of the series, *The Supply–Demand Outlook, 2000–2020*. This SEI report draws heavily on the long-term international energy projections of the U.S. Department of Energy and the Energy Information Administration (EIA). The report focuses on the key energy supply–demand trends and their implications for the geopolitical issues discussed in this study.

Figure 2.1 Predicted Changes in World Consumption of Energy, 2000 and 2020

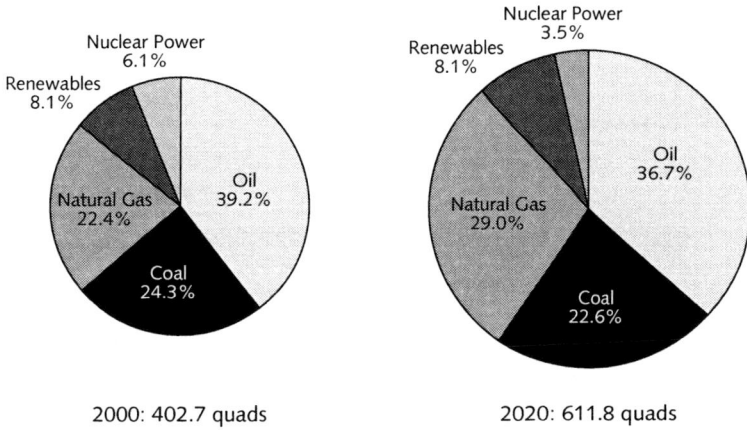

2000: 402.7 quads   2020: 611.8 quads

2020. This expectation of plenty could be damaged, however, should conscious state actions or miscalculations artificially impede supply developments or trade.

Conversely, continuing world prosperity is not necessarily a given. "Asian flu"—the sharp contraction of Asian economies in the mid-1990s—had only a passing but nonetheless sharply depressing influence on world oil prices, forcing exporting nations to set aside personal agendas and support reductions in oil supplies, which in turn led to high oil prices, much to the consternation of U.S. and foreign consumers. Another incidence of Asian flu might not be so easily contained and could bring more lasting changes in the world energy supply-and-demand outlook, with concomitant geopolitical implications.

Given this energy outlook, important changes will take place in terms of the dynamics between exporters and importers, with some changes rather obvious and others less so. The net effect will be to increase the mutual dependencies between energy suppliers and consumers during the first 20 years of the century.

- The Persian Gulf will remain the key marginal supplier of oil to the world market, with Saudi Arabia in the unchallenged lead.

- While the Persian Gulf share of world oil production continues to expand, the share of world oil production from the most stable regions in the Atlantic Basin—the United States and Europe—will decline by one-tenth. Expansion of Canadian oil sales to the United States will help offset a portion of that decline.

- The share of world oil production from the former Soviet Union could increase, from 9 percent to almost 12 percent, but only if timely foreign investment and export pipeline construction are forthcoming.

- Asian dependence on Persian Gulf oil will rise in importance, both in terms of volumes and in terms of political relationships.

- European dependence on Persian Gulf oil will be significant.

- U.S. net oil imports, which broke their 20-year high in 1997, will continue their steady growth.[2]

The expanding role anticipated for natural gas is accompanied by its own set of physical and political vulnerabilities.

- The European need for natural gas will be covered by a handful of suppliers, raising the question of security of supply especially in light of limited alternative transportation options. Russia, for example, in 1999 provided 30 percent of the gas consumed in Europe, a relatively high dependency on a single supplier and likely to grow.

- Anticipated growth in the use of natural gas must be preceded by massive investments in the infrastructure of the natural gas sector, and the need to provide the required infrastructure will be particularly burdensome for developing nations.

In 2000, electricity is the most rapidly growing form of energy consumption. At the close of the twentieth century the consumption of fuel in power generation accounted for 40 percent of total world energy use, up sharply from just under 26 percent in 1973. As would be expected, the overwhelming share of generation occurs in developed countries. But change is coming as developing countries place electrification high on their agenda. Two questions emerge: How should the construction of power plants and distribution grids be funded? What kinds of fuel will be burned in power generation? Not only must the question of fuel identification and supply be addressed but also the impact of fuel choice on the environment.

In light of the continuing and, in some cases, increasing reliance on fossil fuels, governments confront the highly contentious issue of environmental degradation and energy use. The growing politicization of the environmental debate, particularly regarding climate change, will make it increasingly difficult to balance the need for sound environmental policies with the need for policies that recognize the central role that energy plays in economic life.

---

2. Import dependence does not necessarily create vulnerability provided that importing nations draw from diverse sources and make provisions for supply interruptions.

CHAPTER 3

Geopolitics and Energy: A Symbiotic Relationship

Geopolitics and energy existed in a symbiotic relationship for most of the twentieth century.[3] There is little doubt that this symbiosis will continue, offering both dangers and opportunities for policymakers, energy producers and suppliers, and energy users. Geopolitical developments will influence outcomes, but which developments? Conversely, how might energy supply and demand carry geopolitical repercussions during the years ahead?

How Might Geopolitics Affect Energy?

Four key geopolitical trends may influence significantly the availability and reliability of energy supplies, levels of world energy demand, and the ways in which energy is both produced and consumed. These are (1) the behavior of certain major world powers in the wake of the Cold War, (2) the continuing political fragility of many key energy-producing states, (3) economic globalization, and (4) the growing influence of NGOs.

World Powers and Conflict

The end of the Cold War and the ongoing search by the world's great powers for their appropriate global or regional roles have changed fundamentally the geopolitical context in which world energy markets operate. As these great powers jockey for position in pursuit of their national interests and political objectives, they may affect, deliberately or unintentionally, the availability and reliability of energy supplies.

In essence, the world has shifted from a bipolar balance of power to a world where only one country, the United States, possesses all of the attributes—political, military, economic, and cultural—that constitute a true superpower. However, as Samuel Huntington observed in his 1996 book, *The Clash of Civilizations,* this is not a unipolar world. First, a number of other countries qualify in their own right as either regional or global powers: Russia, China, India, and Brazil, not to mention a unified Europe as a regional and global power. In addition, the United States seems unlikely to act in a consistently preeminent manner on the world stage during the period 2000–2020. Instead, U.S. foreign policy will face three sometimes

---

3. For a detailed analysis upon which these conclusions are based, see the companion SEI report, volume 3 of the series, *The Geopolitical Outlook, 2000–2020.*

contradictory sets of policy challenges: the management of its existing alliance relationships; the need to respond to unforeseen international crises; and the pursuit of stated U.S. global priorities such as democratization, the promotion of U.S. economic interests, and the desire to control the spread of weapons of mass destruction (WMD).

Whether or not the United States exercises its power in a limited or more expansive manner, U.S. foreign policy choices in each of these areas will strongly influence energy supply. For example:

- The United States will need to manage a rebalancing in its security relations with its European allies as the Atlantic Alliance tries to coexist with an emerging autonomous European security and defense identity. U.S. and European leaders may find that their respective security priorities diverge, leaving more scope for differences in objectives and tactics and a greater risk of uncoordinated or conflicting responses to crises in energy-sensitive regions such as the Middle East and the Caspian.

- Former representative Lee Hamilton, who chaired the House Foreign Affairs Committee, has stated: "Ensuring access to Middle Eastern energy resources is a vital national interest of the United States."[4] The extent of U.S. engagement in the Middle East, whether in the Arab-Israeli peace process or as an ally to key Persian Gulf monarchies, will be a critical factor during 2000–2020 and beyond, as it was during the last three decades of the twentieth century, in ensuring uninterrupted exports of crude oil from the Persian Gulf region.

- The future viability of the newly independent states of the Caspian region and Central Asia as major energy exporters will be circumscribed not only by Russia but also by the foreign policy priorities of the United States. The United States will seek to balance its interests in nurturing the survival of these countries as sovereign states while it maintains separate and distinct policies toward Turkey, Russia, and Iran.

- U.S. security commitments to Japan, South Korea, and Taiwan provide the geopolitical backdrop for U.S. naval protection of the sea lanes and choke points through which oil from the Persian Gulf must pass in order to reach its many destinations in Southeast Asia and the Far East. Any weakening of these security commitments now or in the future would carry with it increased concerns about the reliability of oil supplies and a concurrent effort by countries in the region to replace the U.S. presence.

- U.S. concerns about the proliferation of WMD will, in all probability, continue to act as a powerful variable in the geopolitical context of energy availability. Major energy exporters, such as Iraq, Iran, and Libya, that also pursue active programs to acquire these capabilities will likely find the United States to be an implacable opponent of their full integration into world energy markets.

---

4. Lee Hamilton, "Can Today's Rogue States Be Tomorrow's Key Energy Suppliers?" (speech at the CSIS conference, "The Geopolitics of Energy into the 21st Century," Washington, D.C., December 9, 1999).

- The tendency of the United States to want to promote the process of democratization around the world, however selectively, could also continue to act as a powerful influence on energy availability. Energy exporters whose governments appear to contravene the basic principles of democratic pluralism may become targets of new sanctions initiatives similar to those imposed on Myanmar.

Even as the United States struggles with the burdens and responsibilities of its position as the world's sole superpower, the actions of other countries can also have a significant impact on the availability and/or reliability of energy supply around the world. Enmities abound, in some cases in areas such as the Persian Gulf that are critical to global energy supply. For example, Iran's relationship with Iraq and with some of the smaller Persian Gulf states could deteriorate; or Iraq itself may become the aggressor once again.

Nor can the absence of conflict be taken for granted in other energy-producing regions—Colombia and Venezuela, for example. In both the Persian Gulf and Northeast Asia, and even in non-energy-producing regions, the likely proliferation of not only WMD but also increasingly destructive conventional weapons such as long-range missiles means that the fallout from any interstate conflict may be longer lasting as well as more immediately damaging in terms of energy supply interruptions. A military confrontation involving China and Taiwan or North and South Korea would have an impact on patterns of energy flows that would expand throughout the region the effects of any conflict. China, in particular, may emerge early this century as a regional power in Asia and one that may seek to challenge the U.S. role as the region's primary military protector.

Apart from the United States, however, the country whose geopolitical ambitions may have the broadest impact on energy is Russia. Russia will remain a major Eurasian power, a power that has long seen energy as not only a factor in its foreign policy but more so an earner of much-needed hard currency. In the Caspian region, the Russian government early on provided the newly independent Caspian states with only limited access to existing Russian pipelines in an effort to slow down their drive for independence. Then, as Azerbaijan, Kazakhstan, and Turkmenistan have begun to expand their output of crude oil and natural gas, Russia has become more accommodating and willing to provide outlets to demonstrate, first, that Russia is indeed a reliable partner and, second, to keep as much of the oil and gas as possible transiting Russian territory.

North of the Caspian region, the Russian government has not been averse to using its position as Ukraine's principal energy supplier to impose other aspects of its foreign policy agenda on Ukraine. In sum, Russia's geographical position, its abundant energy resources, and its military strength all ensure that it will be a major geopolitical influence on the evolution of energy markets throughout Eurasia.

Political Instability among Key Energy Suppliers

A second geopolitical trend affecting the availability and reliability of energy supplies at the beginning of the twenty-first century is the fragility of many of the world's energy-producing states. In one sense this is nothing new. The world had been drawing energy supplies from unstable countries and regions during much of the twentieth century. This is no reason for complacency, however. As noted on page 4, from 2000 to 2020, demand by developing countries for oil and gas imports from the world's current energy suppliers will double. New regions, developing Asia in particular, will become as dependent on energy imports as are the United States, Japan, and countries of Western Europe in 2000. Equally important, by 2020, fully 50 percent of estimated total global oil demand is expected to be met by exports from countries with sensitive or high-risk production capacity.

Moreover, in addition to the perennial unpredictability of many supplemental oil producers, the internal stability of key major energy producers may be increasing in unpredictability. For example, Saudi Arabia and Russia, the world's two leading oil producers, face totally different but equally pressing problems.

- Saudi Arabia is expected to maintain its position as the world's largest oil exporter through 2020. Despite warnings for many years about its internal fragility, the Saudi political system has been stable since its inception in the 1930s. And despite the continuing overdependence of its economy on the oil industry, Saudi Arabia has solidified its position as a highly reliable and leading supplier of crude oil to world markets. Yet, the challenges it faces at the turn of the century are of a different nature from those of the past. Burgeoning population growth, a lack of economic diversification, and an educational system that is not responsive to the needs of a modern global economy have led to high levels of actual and hidden unemployment. Meeting popular expectations will be an increasingly complex challenge for the kingdom's rulers.

- Russia is expected to retain its position in the next decade as the world's second largest supplier of oil if the required foreign investment is forthcoming, and it will also remain the world's largest trader of gas. Despite the emergence of a middle class and an entrepreneurial class in certain major cities, the vast majority of Russia's people remain mired in poverty while a very small minority has become very wealthy. Vast sectors of its economy are now controlled by organized crime, the legal framework for business is still in its infancy, and a viable financial system has yet to take root. The weakness of Russia's parliament, Duma, and a reluctance of the presidency to assert itself had been conspiring to drain power toward increasingly assertive regional governments, a situation that President Vladimir Putin is moving to correct in 2000. Russia faces three possible scenarios: internal implosion and anarchy; the resurgence of nationalism, fostered by an authoritarian leader; or a continuation of its very gradual transition to a more market-based and democratic society.

  Even under this last scenario, the circumstances necessary to attract significant foreign investment to rehabilitate and expand the oil and gas sectors will be very hard to achieve. At the same time, investment can and would be made under an authoritarian leader if investors were assured that their prospects would be fully protected under the law and that these laws would be enforced.

Several other states upon whom energy importers depend and whose national income in turn is heavily dependent on energy exports include Iran, Iraq, Libya,

Algeria, Venezuela, Nigeria, and Angola. These states are politically unstable in 2000 and will find great difficulty in pushing through the necessary internal political and economic reforms to meet their expected energy production targets and, just as important, hold domestic upheaval at bay. Together they share the characteristics of petro-states, whereby their extreme dependence on income from energy exports distorts their political and economic institutions, centralizes wealth in the hands of the state, and makes each country's leaders less resilient in dealing with change but provides them with sufficient resources to hope to stave off necessary reforms indefinitely or until change is forced because of an internal crisis. There is, therefore, a significant risk that a crisis in one or more of these key energy-producing countries could occur during the span of the 20 years at the beginning of the century. Consequently, ensuring broad diversity among sources of oil and gas identified and developed around the world will help reduce the risk of dependency and should be a focal point of U.S. policy.

Globalization

Price of oil, not origin of oil, matters most to consumers and, because it does, actions by policymakers tend to be narrow and reactive, not proactive. What immediate measures might be taken to help mitigate price increases? Unfortunately, because of the heavy U.S. reliance on foreign oil imports—reaching 56 percent in 2000 and scheduled to increase roughly in parallel with demand—most pricing power rests in the hands of oil exporting countries, whose national interests do not necessarily coincide with the importers' interests.

The process of economic globalization is producing a number of effects that are relevant from an energy standpoint. U.S. national energy policy also continues its unfortunate inward focus, a trend that must change to a more international outlook if it is to keep pace with the changing times.

First, as was apparent during the Asian financial crisis, economic collapse in one part of the world can easily and quickly spread to other parts of the global economy. The effects are evident not only in financial markets and economic activity but also in energy consumption. Part of the reason for the collapse in oil prices beginning in 1997 was the financial crisis–induced dramatic fall in oil demand in Asia, which was concurrent with both an OPEC decision to raise output levels and the return of Iraqi oil to the world market. Rapid and unexpected swings in energy demand caused by global economic minirecessions and recoveries could become an occasional feature of the energy market of the twenty-first century.

Conversely, the global spread of market-based economics in the decade after the end of the Cold War imposed new pressures on the energy markets just as it did on all forms of domestic and international economic activity. The integration of national economies into a web of trade, foreign direct investment, international financial flows, and technological innovations is breaking down the barriers between countries and companies. In energy, national champions are being supplanted by international corporate brands.

The process of deregulation and privatization that has resulted from globalization is transforming energy markets. The growth of gas as a power source, for example, has been a direct result of this process (see figure 3.1). For their part,

Figure 3.1. Predicted Changes in World Generation of Electricity, 2000 and 2020

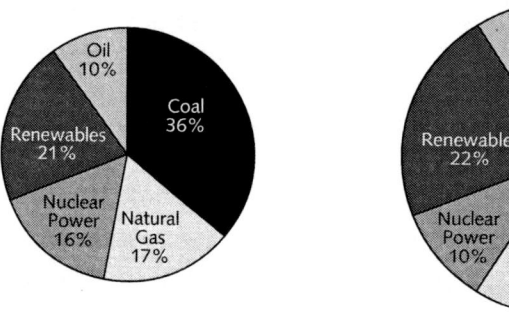

2000: 153.9 quads   2020: 227.0 quads

online markets hold the promise of allocating energy products with near-perfect efficiency. If the globalization of the world economy, the boom in both energy and information technologies, and the resulting changes in the structure of the energy business continue, we may witness, according to Peter I. Bijur, chief executive officer of Texaco, a "resistant world energy market not easily pinched or shut down by geopolitics."[5]

It is worth noting that in an age when the pace of technological change is almost overwhelming, the world will essentially remain dependent, during 2000–2020 at least, on the same sources of energy—oil, coal, and natural gas—that dominated during the last century. But the world itself will change as it transits to a new economy, a new economy based in considerable part on the spread of information technology and an expanding reliance on the Internet. This new economy in turn will be increasingly dependent on reliable supplies of electricity. Such dependence unfortunately carries a set of vulnerabilities with it: exposure to terrorist attacks on the physical electrical infrastructure and to cyberterrorist attacks on the operating infrastructure.

The Rise of the Nonstate Actors

Globalization, the spread of information technology, and use of the Internet are making a dramatic impact not only on economic structures and performance. These trends are also changing the ways in which political power is exercised. In traditional political structures, governments set policy agendas and pass laws as part of a complex but relatively closed triad involving legislators, civil servants, and special interest groups from the private sector. Today, well-organized and financed groups of citizen-activists are addressing with increasing impact those policy issues that formerly were the sole province of national political structures or of treaty-based international organizations.[6]

---

5. Peter I. Bijur, "A New Era for Energy Suppliers: Challenges and Opportunities" (speech at the CSIS conference, "The Geopolitics of Energy into the 21st Century," Washington, D.C., December 9, 1999).

6. Jessica T. Mathews, "Power Shift," *Foreign Affairs* 76, no. 1 (January/February 1997), 50.

One of the primary reasons for the dramatic growth of influence enjoyed by NGOs is not only their abilities to tap into growing public concerns but also their creative use of the tools available through the expanding information revolution. Concurrent with the end of the Cold War and the loss of a consensus approach to foreign policy in many of the world's democracies, nongovernmental advocacy groups have taken advantage of the explosion in cheap, accessible communications, including the Internet, and of the growing influence of the visual media. These groups can now communicate their views to large audiences, organize and mobilize support not only domestically but also across national borders, and engage public opinion directly into the political process around a single issue.

The Greenpeace environmental campaign against the disposal at sea of the *Brent Spar* oil storage platform and the human rights campaigns against the military regimes of Myanmar and Nigeria are but two examples of the way that governments are increasingly pressured by NGOs to take decisions that impinge directly or indirectly on energy production. It is possible that public attention could be similarly aroused on future issues, such as women's rights or the lack of democratic institutions in major Persian Gulf oil-exporting countries. About 40 percent of the world's oil supply is produced in countries that, as of 1999, had yet to sign or ratify the main United Nations (UN) human rights covenants or that are subject to serious criticism by the U.S. State Department and human rights organizations for inadequate protection of human rights.

Nongovernmental groups of a different sort are also exercising a growing impact on the reliability of energy supplies. Terrorist groups have often targeted vulnerable oil and gas pipelines (for example, in Colombia and Algeria) as a cheap but highly visual way to express their opposition to government authorities or to foreign company operations. The Internet, encryption, and satellite communications now accentuate further the built-in advantages that such small groups can have in their struggle against large hierarchical and often poorly funded security forces in energy-producing countries. As these energy-exporting countries continue to expand their production capacities and as gas pipelines play an increasingly important part in energy delivery, the vulnerability of complex energy infrastructures to various forms of terrorist attack, including cyberterrorism, will be a growing preoccupation for governments around the world.

How Might Energy Affect Geopolitics?

Although the potential impact of geopolitical forces on the reliability and availability of energy supplies is relatively well appreciated, the reverse relationship has tended to receive less attention. Over the next two decades, however, it is likely that the effects of energy on geopolitical outcomes will be of increased salience. The four principal ways in which this interaction will likely be most apparent are (1) the effect of swings in energy demand upon key energy-producing countries, (2) competition for energy in Asia, (3) the potentially integrating effect of energy relationships around the world, and (4) the struggle among policymakers for making either energy or the environment the primary consideration.

Swings in Energy Demand

The growing economic integration of countries around the world as a result of the process of globalization may lead to significant swings in energy demand on occasion during the next two decades. The excessive price volatility that may accompany such swings would have negative effects for both producers and consumers. For many of the world's energy-exporting countries, a substantial reduction in oil and gas export revenues caused by a significant decline in prices deriving from reduced global energy consumption could trigger the very internal political instability that has been described. Most energy-exporting countries are still carrying heavy external debt burdens incurred when they were unable, or perhaps unwilling, to diversify their economies and when funds had to be borrowed to maintain their generous social welfare programs. Conversely, for consumers, sustained rises in price could reignite inflationary pressures; and for multinational energy companies, price volatility is a disincentive to investment.

Clearly, countries such as Russia, whose 1998–1999 financial crisis was greatly exacerbated by—and possibly precipitated by—reduced hard currency revenue generated by its energy exports, and Saudi Arabia, where foreign investment and diversification plans hinge on sufficiently high oil prices, and Venezuela, where the population expects significant economic improvement under its new government, have the most to gain or the most to lose by fluctuating oil prices. In a worst-case scenario, internal strife in one or more major energy exporters as a result of a global recession could lead to a fall in oil exports and a decline in production capacity that could not then be immediately reversed as world economic growth would begin to pick up speed again. Looking ahead, it appears that the starkly conflicting relationship between producing and consuming countries that characterized the 1970s is being supplanted in the early twenty-first century by a recognition of each other's interests and vulnerabilities and a growing mutual interdependence.

Competition for Energy Supplies in Asia

In contrast with other geographic areas of the world that have a much more favorable combination of indigenous energy resources and diversity of import alternatives, the rapidly industrializing countries of Asia are unusually poor in indigenous energy sources and commensurately dependent on long-distance imports of oil and natural gas.

Because they are keenly aware of their dependence on external energy supplies and are driven toward the completion of an industrialization process that is crucially intertwined with energy use, individual Asian countries place energy security high on their lists of foreign policy concerns. These countries are in an inherent situation of energy competition on a regional basis, and this situation can only be expected to intensify should a sudden scarcity of world energy supplies occur. During 2000–2020, China, South Korea, and India are expected to increase very substantially their respective shares of Asian oil imports. By 2020, China's share of Asian oil imports might equal the share that Japan has today.

Future energy competition in Asia will not take place in a geopolitical vacuum. Tensions and enmities in Asia have deep historical roots, and rivalries are inherent in the geography of the area. Japan still has troubled bilateral relationships with China, Korea, and Russia dating back to its wars against each of these countries in the twentieth century. India and China have evolved into geopolitical rivals and have a history that includes border warfare. China and Russia are also geopolitical rivals, with their history of border clashes and ideological competition.

Competition for energy resources, therefore, carries a potential for escalation into armed conflict in future years. The most obvious example of this interlinkage concerns the competing territorial claims of the Southeast Asian countries in the South China Sea, where the potential for oil discovery is believed to be high. China has asserted a historical claim to sovereignty over almost the entire South China Sea, a claim that has already led to both diplomatic and actual skirmishes over the Paracel and Spratly Islands. Moreover, these disputes could disrupt oil tanker traffic in transit from the Persian Gulf to Asian markets.

A different geopolitical implication of China's search for reliable energy supplies may surface in a building of closer ties between China and the oil-exporting countries of the Persian Gulf and, later, of the Caspian basin and of Russia itself. Building these ties could extend to a tightening of military cooperation and military sales from China to Persian Gulf states such as Iran and Iraq. That trend would again reinforce diplomatic and weapons-proliferation concerns on the part of the United States.[7] In addition, proposed exports of crude oil and natural gas from Russia to China could be expected to lead to closer political relationships between these two countries, with unpredictable consequences for their relations with the United States, Europe, and their Asian neighbors.

Energy and Regional Integration

Although it is easy to define the risks inherent in the various linkages between energy and geopolitics, it is also important to note that energy could serve to strengthen and stabilize bilateral and regional ties in a number of instances during the coming years. The rise in the use of natural gas, driven partly by the application of combined-cycle technology, is central to this process. Whereas the rise of nuclear power in the 1970s reflected a search for autarky and energy security, use of natural gas is integrative. Not only does it cut across boundaries, it ties economies together. At the same time, natural gas cannot yet be considered a global commodity; it remains regional in scope.

Energy distribution networks in Latin America could play a vital role not only in regional economic growth but also in helping overcome persistent border disputes between the trading partners of the Mercosur (Mercado Común del Sur or Common Market of the South) grouping. Brazilian consumers already are using electricity provided by Uruguayan power plants that are reliant on Argentine natural gas. Clean hydroelectric power will soon be sold by Venezuela to Brazil's northern states. Peru's Camisea gas field, the largest in Latin America, is attracting U.S., South American, and European investors. Bolivia is quickly becoming an energy hub. In Central America, European, Asian, and U.S. companies are explor-

---

7. Others may point out that a China dependent on oil imports from the Persian Gulf would confront a vulnerability that in turn plays to U.S. interests.

ing innovative solutions to providing electricity between communities and across borders known in the past more for political conflict than peaceful and economic cooperation. "[T]he fusion of energy interest is really driving the political and economic integration in the Western Hemisphere because when you rely on energy from your neighbors you are not linked just to their power grid, but in many ways you are linked to their prosperity."[8] In most of these cases, it is the private sector that is driving the integration process.

Even though the distances are great and the intervening terrain often difficult, a natural complementarity exists between the large reserves of oil and gas in eastern Russia and Central Asia and the enormous pent-up demand in Asia. "Relations between Russia and China, instead of being based on Marx and Lenin, will be based, in the future, on oil and gas."[9] Not only might China try to find ways to import energy directly for its own use, it could also serve as a land bridge for energy transportation to Japan and Korea. Farther south, Asia-Pacific Economic Cooperation (APEC) member-countries are already outlining plans for the construction of gas transmission and distribution networks and electrical grids linking consumers throughout Asia. Nonetheless, neither the technical difficulties nor the political hurdles to completing energy integration in this region should be underestimated. Yet energy networks, along with a further liberalization of trade in the region, could serve to reduce historical tensions and place Asian economic growth on a firmer footing for the future.

A second instance of energy serving to link together geopolitical rivals lies in Europe. Western European states and many of their central European neighbors already import large volumes of gas from Russia. While this growing interdependence was a source of great concern to the United States during the Cold War, it offers one of the few opportunities in which the EU can build concrete ties with Russia in the early years of the twenty-first century, given the parlous state of other sectors of the Russian economy and the difficulty of striking any meaningful trade agreements. The risks of EU dependency on Russian gas could be mitigated by Europe's continuing to secure diverse sources of supply and by ensuring that other sources of natural gas continue to be available—from Norway and Algeria, for example—to help compensate for possible interruptions in the Russian supply. Unfortunately, most gas deliveries are tied into long-term sales contracts and there are few prospects for diversion, even in small volumes, to other markets.

Finally, Bangladesh may overcome its reflexive suspicions of its large neighbor and start exporting natural gas to India early in this century.

Energy, the Environment...and Geopolitics

One of the more radically changing aspects of the worldwide energy situation at the beginning of the twenty-first century is the extent to which concerns about the environment have become intertwined with debates about energy supply and use.

---

8. Thomas F. McLarty, "Energy Integration in Latin America" (speech at the CSIS conference, "The Geopolitics of Energy into the 21st Century," Washington, D.C., December 9, 1999).

9. Daniel Yergin, "Geopolitical Risks in the World Energy Market" (speech at the CSIS conference, "The Geopolitics of Energy into the 21st Century," Washington D.C., December 9, 1999).

Figure 3.2. Growth of Energy Consumption, 2000–2020

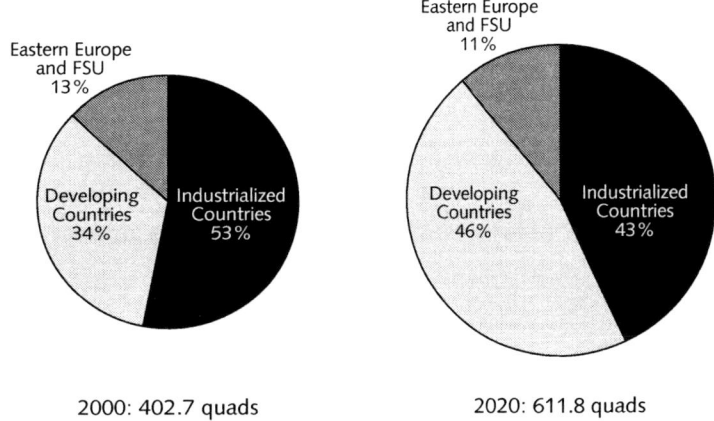

2000: 402.7 quads   2020: 611.8 quads

Perhaps the most dramatic shift was the perception during the final decades of the twentieth century of the borderless nature of environmental damage, highlighted by the rising concern over acid rain, depletion of the ozone layer, marine pollution, and, most recently, climate change.

Given the many scientific studies that have argued—for and against—the proposition that man-made carbon emissions are a leading source of global warming, energy use lies at the crux of the current debate about the development of appropriate international measures to deal with climate change. Governments in the industrialized world are under pressure to confront the issue, partly because of a precautionary desire to cope now with a phenomenon that, if certain projections are borne out, could impose considerable human and financial costs on governments in the future, and partly because of increased public pressure, mobilized by a plethora of interested NGOs.

Yet there are also scientific studies that argue that the causes of climate change and global warming are still subject to confirmation. The ensuing debate over the real causes is often highly contentious, which complicates the issues for government and for corporations.

Appropriate response to climate change is indeed a quintessentially global issue and will require the utmost level of international cooperation for implementation. Because the process of industrialization is so crucially linked to those factors believed to cause climate change, it is not surprising that sharp divisions have emerged between the interests of advanced countries, which have already industrialized, and those that are currently in the process of industrializing. The Kyoto Protocol reflects this difference by permitting the developing countries to defer a decision-binding commitment to reduce their own carbon emissions until the developed countries have undertaken effective emissions-reduction measures themselves.

Developing countries, where energy consumption is projected to exceed that of developed countries during 2000–2020 (figure 3.2), are reluctant to undertake any measures that might retard their economic development, despite concern about specific environmental issues such as urban pollution. These countries argue that the immediate obligation to reduce current and projected levels of carbon emissions rests with the industrialized countries because it is the already completed process of industrialization that accounts for the high level of carbon and other greenhouse gases in the atmosphere. In addition, on a per capita basis, industrialized countries produce far more atmospheric carbon than developing countries. For example, China's per capita emissions are today one-ninth of the per capita emission in the United States. That implied advantage is, of course, offset by China's huge population, some five times that of the United States.

Ratification of the Kyoto Protocol has encountered considerable opposition, however, especially in the United States, where the protocol's focus on near-term mandatory reductions is perceived by many to impose particularly onerous costs on the United States. In addition, the counterargument is made that over time far greater quantities of carbon will be contributed by developing countries—first, because they have large and rapidly growing populations and, second, because they will be constantly increasing their own levels of industrialization. For example, by 2020, as a result of its continued dependence on coal for electricity generation, China's total level of carbon emissions is projected to equal that of the United States. Indeed, by 2020, developed countries might come to see the developing world's growing contribution to global warming as threatening their security. Conversely, to the extent that the negative effects of global warming are evident first in the developing world, developing countries may perceive that the blame lies with the refusal of the developed world to pay for its earlier actions.

The potential is clear for climate change and energy use to create acute fault lines between the developed and developing worlds as well as within the family of developed countries because they might perceive different levels of urgency for action. In 2000, how these fault lines will be bridged is unknown. No one knows yet who will be the relative losers; and no one knows when and who, if anyone, might gain from climate change. It is still a matter of scientific debate and conjecture.

CHAPTER 4

Energy and Geopolitics: Policy Considerations

It is the central premise of this study that policymakers should promote energy supplies that are adequate to meet the demands of a robust global economy, reliable enough to avoid interruptions damaging to the global economy, and produced and consumed in ways that protect the environment.

The evolution of geopolitics in the first 20 years of the twenty-first century can potentially affect each of these energy goals dramatically. Although there is no overall shortage of energy resources on the planet, geopolitics can constrain their availability. Both the relatively small number of major energy suppliers and the vulnerability of land and sea transportation routes offer opportunities for unwanted supply disruptions. The ability to produce and transport energy supplies in environmentally sound ways depends in large measure on the outcome of a North–South dialogue that has only just begun.

Against the background of key outlines of the world's energy outlook between 2000 and 2020 and in view of the full spectrum of potential implications that geopolitical developments may hold for that energy outlook, what are the most pressing challenges to policymakers? In what ways are existing foreign and domestic policies working at cross-purposes to or in support of the need for adequate and reliable supplies of energy that are produced, delivered, and consumed in an environmentally acceptable manner?

Potential policy considerations may be formulated along the three lines of this central premise. How can policymakers ensure that energy supplies will be adequate to meet the world's energy needs in the longer term? What steps might help ensure the reliability of energy supplies and avoid any major interruptions in the short term? What steps could policymakers realistically be pursuing to ensure that energy supply and demand are satisfied in a practical, environmentally acceptable manner?

Energy Availability

Fossil fuels will continue to meet the bulk of energy requirements for at least the first 20 years of the twenty-first century. No shortage of reserves is expected for any of the three principal fossil fuels—crude oil, coal, and natural gas—during 2000–2020. Unfortunately, these reserves are often remote from consuming centers, and prevailing political and infrastructural constraints limit deliverability.

Plentiful energy to sustain global economic growth depends on developing these reserves and transporting the fuels on a timely basis in the needed amounts to the waiting markets. In the first two decades of the twenty-first century, the world's dependence on a few suppliers to meet its growing demand for petroleum will increase. Moreover, these suppliers will continue to be concentrated primarily in volatile or potentially unstable regions of the world. A principal dilemma for Western governments, therefore, is that ways must be found to access the production potential of three of those states—Iran, Iraq, and Libya—that in 2000 have either bilateral or international sanctions imposed on them.

The Contradiction of Sanctions...

Western policy today contains a fundamental contradiction. Projections indicate that oil and gas exports from states experiencing sanctions will play an increasingly important role in meeting growing global demand, especially in avoiding increasing competition for energy resources with and within Asia. However, U.S. sanctions on Iran and Libya and UN sanctions on Iraq have constrained the development of those states' energy resources and may prevent the timely construction of the infrastructure necessary to meet the growth in energy demand. Moreover, these sanctions have increased Western dependence on Saudi Arabia, Kuwait, and the United Arab Emirates while they have strained relations between the United States and its principal European allies.[10]

Governments should avoid indiscriminate use of sanctions in general, especially toward major energy-producing countries. Governments should adopt a graduated approach and make every effort to ensure that sanctions, when they are called for, are as narrowly targeted as possible. The decision should include a cost–benefit analysis of whether curtailing investment in, or revenue from, energy production will genuinely dissuade the target government from the specific behavior that provoked the imposition of sanctions. At the same time, engagement and dialogue should not be abandoned.

...and of Politics over Economics

A similar contradiction exists in U.S. policy toward the Caspian region and Central Asia. The United States is committed to reinforcing the political and financial independence of these former Soviet republics in what Zbigniew Brzezinski has called the "arc of crisis." Development of reliable revenues from oil and gas exports will be central to this objective, but U.S. policymakers have expressly maneuvered to secure export routes that would bypass both Iran and Russia.

---

10. For a thorough analysis of the use of sanctions as an instrument of foreign policy and a set of recommendations to guide future use of sanctions, see *Altering U.S. Sanctions Policy: Final Report of the CSIS Project on Unilateral Economic Sanctions* (Washington, D.C.: CSIS, 1999); Joseph J. Collins and Gabrielle D. Bowdoin, *Beyond Unilateral Economic Sanctions: Better Alternatives for U.S. Foreign Policy* (Washington, D.C.: CSIS, 1999); and Ernest H. Preeg, *Feeling Good or Doing Good with Sanctions: Unilateral Economic Sanctions and the U.S. National Interest* (Washington, D.C.: CSIS, 1999).

The United States has insisted on an oil export route from Baku to Turkey although the route's economic feasibility has been questioned because sufficient volumes of crude oil to justify its construction are lacking. Final pipeline costs are an additional concern. Apart from the desire to support Turkey, a strategic ally for the United States, U.S. policy also reflects its determination to forestall the construction of a pipeline network from the Caspian through Iran that would provide a new revenue stream for Iran and make it a central player in the region's future.

Oil and gas exports from the Caspian and Central Asia hold the prospect of becoming a valuable and diverse source of energy supply. Even as the U.S. government endeavors to make feasible an East–West transportation corridor that bypasses Russia and Iran, it should not obstruct the development of alternative routes that would ultimately offer exporters in the Caspian and Central Asia choices based more on geography and economics and less on politics.

The opening of the former Soviet Union to foreign oil investment and subsequent successes in oil exploration in the Caspian Sea triggered expectations, fueled by the media, that at long last a substitute for Middle East oil had been found. But realities soon set in with the recognition that, although Caspian and Central Asian oil will be important to the world market at the margin, these new supplies will not be pivotal.

Another major policy challenge for increasing supplies of fossil fuels in the first two decades of the twenty-first century falls more in the realm of geoeconomics than geopolitics. The new crude oil and natural gas reserves that must be developed to meet this century's demand are found, for the most part, in developing nations that have yet to build viable market economies. Massive foreign investment will be required. Even though international financial institutions can provide a portion of the needed capital, these requirements dwarf the ability of the public sector alone to respond. Thus most of the burden of financial support will fall necessarily to the private sector.

Opening Up to Foreign Investment

Western governments should make use of whatever political leverage they possess, including openings for foreign investment, within key energy-exporting countries to encourage market reforms that will improve the performance of the energy sector. This leverage would include provisions for the enforcement of contracts, guarantees for private property, anticorruption measures, and stable tax regimes. Attracting foreign investment is by no means automatic, however. Foreign investors will weigh carefully the openness of the markets of the emerging nations, the extent to which the rule of law applies, and the degree to which foreign investors can operate freely within the country.

Foreign investors encounter resistance in the emerging nations for a variety of reasons. Opening borders to the vagaries of globalization can mean subjecting the national economy to the tyranny of unfettered market forces and to volatile boom-or-bust cycles. Moreover, adapting formerly directed economies to market conditions invariably produces wrenching adjustments that in turn can provoke social unrest and, possibly, political instability. The political leadership in many emerging nations remains sensitive to charges of subjecting the country to neocolonialism if foreign investors are given control of large stakes in key industries, particularly those related to energy. Natural resources are regarded as a national birthright, and turning these resources over to outsiders can provoke vehement opposition. The policy challenge facing Western governments will be to encourage the key energy-producing countries, especially in the Persian Gulf, to continue to open their energy sectors to greater foreign investment. This approach must be sensitive to the potential for nationalistic backlash, and success will come only if these exporting countries believe that opening up will be in their own national interests.

The interplay of geopolitical and geoeconomic forces is likely to be highlighted in Asia, where the policy challenges are especially pressing. The huge potential demand for natural gas in the two most populous countries in the world—China and India—over the first two decades of the twenty-first century can be satisfied only by pipeline links with remote gas fields in Central Asia and/or Russian East Siberia and the Far East. The costs for such pipelines are enormous because of both distance and the difficult terrain that must be traversed. But China, for example, may judge that diversity and security of supply justify the costs.

It is in the interests of the United States and other Western governments to support China's efforts to seek diversification among kinds of energy consumed and sources of energy imports, avoiding excess dependence on any one fuel or any single supplier. At the same time, the resultant political linkages might not serve U.S. interests.

Tensions in South Asia already restrict the ability of Bangladesh to supply natural gas to neighboring India. The civil war and political isolation of Afghanistan continue to impede the construction of pipelines for bringing gas from Turkmenistan to Pakistan and India and oil through Pakistan to the Arabian Sea. Iran, possessing the second largest gas reserves in the world, is constrained in its ability to export, in part by virtue of sanctions imposed by the United States that have worked to limit gains in producing capacity and constrain construction of export pipelines.

Once again, foreign policy and energy policy have come squarely into conflict. Attracting the necessary investment to build new energy infrastructure linking Russia, Central Asia, and Iran with energy-consuming markets, while desirable from a perspective of improved global energy availability, will present major financial and political challenges.

Energy Reliability

While geopolitics may influence the adequacy of the supply of fossil fuels in the opening decades of the twenty-first century, the more immediate impact of geopolitics is likely to be evident in its effect on the reliability of supply. In both the industrialized and the developing worlds, energy supply interruptions and resultant high prices could trigger recessions and inflict significant damage on national economies everywhere. In the early decades of this century, because burgeoning energy demand will be met largely by oil and gas and because the bulk of oil exports will be increasingly concentrated in a small number of suppliers, the risk posed by supply

Table 4.1. Predicted Changes in Oil Production Capacity, 2000–2020

Region	2000 MMbbl/d	2000 %[a]	2010 MMbbl/d	2010 %[a]	2020 MMbbl/d	2020 %[a]	Change in Production Capacity, 2000–2020 %
Persian Gulf	23.9	30	29.6	31	42.2	38	+77
FSU	7.3	9	10.1	11	13.1	12	+79
Africa	2.6	3	3.3	3	5.5	5	+115
United States	9.1	11	9.0	9	8.7	8	–4
North Sea	6.9	9	7.0	7	5.9	5	–4
World	80.4		95.0		112.2		+40

Source: Calculated from data in *International Energy Outlook 1999* (Washington, D.C.: Energy Information Administration [EIA], U.S. Department of Energy [DOE], March 1999), page 201, table D1.

[a] Percentage of total world oil production capacity.

interruptions will be correspondingly greater than it was at the end of the twentieth century.[11]

Domestic turmoil within key energy-producing countries constitutes another threat to the reliability of energy supply. At least 10 of the 14 top oil-exporting countries demonstrate a potential for political instability. The extent to which political instability in any single exporting country could have a severely negative impact on the world oil market depends, in part, on the market share enjoyed by that exporting country at the time of the disruption.[12]

One of the most important characteristics of the world energy supply in the 2000–2020 time frame will be the growing dependence on oil from the Persian Gulf, as underscored in table 4.1. Only six countries in the world possess oil reserves in excess of 75 billion barrels. All are located in the Persian Gulf. Together, these six account for 70 percent of world oil reserves.

The prospect of war in the Persian Gulf region has not disappeared. Five major wars in the Middle East during the past 30 years—the Arab-Israeli wars of 1967 and 1973, the Iran-Iraq war of 1980–1988, Iraq's invasion of Kuwait in August 1990, and the Gulf War to liberate Kuwait in 1991—testify to the continuing danger there.

Of great importance is the vulnerability of sea and land transportation routes. Fourteen million barrels of oil traffic each day, on average, move through the Strait of Hormuz, which connects the Persian Gulf with the Gulf of Oman and the Arabian Sea. By 2020, a projected 38 percent of oil moving in international trade will transit this single choke point. Its closure would provoke a serious interruption of

---

11. While new sources of supply will come on stream from South America, Africa, and the Caspian region, the EIA projects that most of the growth in oil production during 2000–2020 will come from the Persian Gulf.

12. Iraq and Kuwait together enjoyed a 14 percent share of the oil export market when their exports were interrupted by Iraq's invasion of Kuwait and international sanctions were imposed on Iraq as punishment. At that time, other producers, primarily Saudi Arabia, were able to make up the supply shortfall. Conversely, in 1979 Iran enjoyed a 15 percent share of the international oil export market when the revolution there caused a decline in its oil exports of sufficient magnitude and duration to spark a price spike that triggered a worldwide recession.

oil supply. The Strait of Malacca, connecting the Indian Ocean to the South China Sea and the Pacific Ocean, carries more than 8 million barrels daily, a figure that can be expected to rise dramatically during 2000–2020. Other key choke points include Bab el Mandeb between the Red Sea and the Gulf of Aden and the Arabian Sea; the Suez Canal; and the Bosporus. To be sure, it is possible to divert around the Strait of Malacca and the Suez Canal, if necessary. Nonetheless, such diversions add both time and cost to the delivery of supply. The Bosporus, the key exit route for Russian and Caspian oil, deserves special attention because it could easily be closed by a tanker accident.

Protection of Sea Lanes

The protection of sea-lane transportation routes, to ensure the safe passage of oil supplies, has been well recognized and accepted as a U.S. responsibility since the 1960s. Whether the United States can sustain this role is questionable. U.S. forces have become severely overstretched since the end of the Cold War. Operating tempo has become a source of concern and discontent while the high cost of operations and maintenance is taking a large share of the U.S. military procurement budget. Sustaining a global and dominant U.S. presence under these conditions may become increasingly difficult, especially as regional powers improve their own military capabilities. At the same time, public and legislative support for the U.S. role as protector of key global sea lanes may diminish as it becomes apparent that this role, although of strategic importance to the United States, is of greater benefit to countries in Europe and Asia that are economic and even geopolitical competitors.

The United States should maintain its ability to defend open access to the Persian Gulf and international sea lanes.[13] At a time when the administration faces myriad competing demands for military and peacekeeping interventions as well as resource-consuming domestic priorities, this mission should be accorded a high strategic priority. It may call for greater emphasis on, and increased investment in, appropriate military capabilities.

Equally important, U.S. allies in Europe and Asia should be prepared to accept a greater share of the financial cost of protecting energy supply, including sea-lane protection.

While the United States currently serves as the principal protector of the world's sea lanes, there is no comparable provision in place to protect long-distance energy transportation by land. The growing importance of natural gas in meeting the energy needs of this century heightens the concern regarding pipeline vulnerability. Pipelines carrying oil and gas are subject to closure by accident, military action, and sabotage. Terrorist-motivated sabotage has already disrupted oil and gas pipelines in Algeria and Colombia and has shut down that portion of the Baku-Novorossiisk oil pipeline that passes through Chechnya. While most occurrences thus far have had only local impact and damage has been quickly repaired, terrorist-motivated sabotage is expected to present greater challenges in the coming years. Saudi Arabia for one has taken actions to minimize the threat of supply interruption to its exten-

---

13. The ability to defend open access also implies the ability to control access.

sive routes. However, as the world's largest oil exporter, the sheer size and concentration of processing facilities will continue to present targets of opportunity to terrorists and adversaries.

If a template of a 10 to 15 percent export-market share is applied for contingency planning purposes, it is evident that an export disruption in a single top oil-exporting country—for example, Saudi Arabia—or in a combination of key oil-exporting countries—for example, Russia and Venezuela—could provoke price increases of a magnitude to trigger a world recession. A scenario in which a slump in Russian oil exports caused by political and economic dislocations occurs at the same time as a downturn in production in Venezuela caused by prolonged labor agitation is not difficult to envision.

Admittedly, the value that the exporting countries place on oil-derived revenues would tend to keep the oil flowing under most circumstances. However, military conflict or unresolved civic unrest could interfere with these flows. Thus the possibility of an oil supply disruption severe enough to cause a price spike and a worldwide recession cannot be ruled out during 2000–2020. Shut-in producing capacity in countries other than Saudi Arabia is minimal and, if put to use, would fall far short of covering any significant supply shortage. Western governments must be particularly sensitive to the internal political situation in key energy-producing countries over the coming years and hedge against the danger of a sudden supply disruption.

Governments must find new ways to protect critical energy infrastructure. At a government-to-government level, international agreements to protect pipeline systems might have a deterrent effect. Governments must also find ways to work with the private sector to minimize the vulnerability of all energy infrastructures to sabotage or terrorist attack, including cyberterrorism attacks on operating infrastructure.

The more feasible approach over the near to medium term to mitigating the risks of interruptions to gas supply is to encourage importing countries to promote diversity of suppliers and routes. European governments, in particular, should be attentive to security of gas supply, given the comparatively high dependence on imported Russian gas.

Strategic Stocks

Clearly, contingency planning is necessary to both attempt to prevent such supply interruptions and limit their impact if they do occur. One method of preparing to face potential oil supply interruptions is for governments to finance and control strategic stocks that can be drawn upon in the event of supply shortages. When used during a supply interruption, strategic drawdowns have several advantages: they put more oil into the market, which dampens tendencies toward panic and hoarding; and their introduction into the market also gives time for diplomacy to work, or military preparations to be put into place, in order to address the cause of the interruption. To be effective, however, strategic stocks must be used in a timely fashion. The tendency may be to hold back these stocks for use if the supply situation worsens further.

The United States has a strategic petroleum reserve (SPR) and strategic reserves are also held by several other countries. The IEA, which includes nearly all of the OECD countries, requires that each member hold an oil-stock reserve equivalent to 90 days of net imports, and emergency stock-sharing arrangements have been defined and agreed upon. However, such strategic stocks of oil are absent from most other countries. The absence of a strategic petroleum reserve is especially relevant in Asia where oil consumption and dependence on the Middle East have been rising.

Storage of oil in strategic stocks is costly to maintain, and an opportunity cost is paid when these resources are held off the market. It has, for example, been estimated that each barrel of oil in the U.S. SPR has already cost $59 to purchase and store. Developing countries, while requiring strategic reserves as much as other consuming countries, may well be unwilling to incur the expense of holding reserve stocks because of competing, higher-priority budget demands. Moreover, even in industrialized nations more able to afford strategic reserves, governments may be tempted to use the oil for budget relief or price manipulations instead of hold the reserves for their intended purpose. Finally, as many energy-importing countries become increasingly dependent on gas imports, it is important to recognize that, for the foreseeable future, it will be both impractical and prohibitively expensive to hold strategic gas reserves, thus creating a special vulnerability that must be addressed.

However, the likely dangers to energy supply in the period from 2000 to 2020 require that Western governments maintain and, where appropriate, expand government-financed and -controlled strategic petroleum reserves. This could include extending the IEA emergency preparedness program to nonmember major oil importers and supporting regional stockholding initiatives (for example, APEC).

Energy and the Environment

The debate over climate change and the proper relation among economic growth, energy consumption, and the environment will be a focal point for the public and private sectors during the 2000–2020 time period. The differing points of view among developed countries and the conflict of interest between developed and developing countries will potentially raise climate change to an increasingly contentious geopolitical issue. This will certainly be the case if, as seems most likely, the U.S. Senate continues to refuse to ratify the Kyoto Protocol in its present form and, as a result, there is no coordinated international effort to reduce carbon emissions.[14] Even as governments and NGOs continue to debate and negotiate the parameters of a ratifiable protocol and assess the methods for implementation, they must also prepare for the eventuality that no effective agreement can be reached anytime soon. In this event, Western policymakers must have alternative policy options.

---

14. For a discussion of the relative costs and benefits of the Kyoto Protocol, see the companion SEI report, volume 2 of the series, *The Supply-Demand Outlook, 2000–2020*.

As an alternative to Kyoto, the implementation of a comprehensive research and development strategy directed toward developing low carbon or zero carbon emissions and carbon sequestration technologies deserves serious consideration. This strategy would embrace technologies that include hydrogen fuel cells, advanced nuclear power, and renewable forms of energy.

Fuel Choices

Choices that developing countries make about the way in which they will generate electricity will be crucial in determining not only the extent of local pollution but also, increasingly, in determining the dimensions of the global problem of reducing levels of carbon in the atmosphere and the associated likelihood of climate change. The natural tendency of most developing countries will be to use coal to generate electricity because it is often their most plentiful source of fuel; this is especially true for both China and India. The United States also uses coal as its primary means of generating electricity, and in a number of key consuming countries the coal industry is subsidized or otherwise protected because of its high employment potential and the strength of the industry as a political lobby.

These policies undermine any arguments that developing countries should work to reduce their future reliance on coal.[15] On a worldwide basis, coal use for electricity generation will increase but its relative share is expected to be reduced from a 1999 level of 37 percent of all electrical power generated to a still impressive level of 35 percent by 2020. A central policy challenge, therefore, will be to minimize the negative effects of continued coal use for electricity generation, particularly in the developing world.

The problem of carbon and pollutant emissions from coal-fired electricity plants can be mitigated by the application of clean-coal technology. But this technology is expensive and beyond the financial reach of most developing countries. The multilateral lending agencies should promote guidelines that encourage the use of clean-coal technologies and natural gas. As an additional part of this process, OECD governments should support bilateral agreements with developing countries for joint reductions in greenhouse-gas emissions. These agreements also could include a grant-aid component that would support institution building and the transfer of technical knowledge to the developing countries.

---

15. The impact of reduced coal use on the environment is clear. Indeed, in retrospect, the gain in oil prices during 1999 may not have been the most significant energy event of that year. Instead it is possible that the dramatic fall in the production and consumption of coal in China may have held broader implications, particularly if the declining trend that began in 1997 continues. Because the production and consumption of coal in China is massive, accounting for roughly 25 percent of the world total, any shift is significant. See *BP Amoco Statistical Review of World Energy (2000)*, 32–33, for data that underscore that Chinese coal production declined by 18.1 percent in 1999 while consumption fell by an estimated 16.8 percent.

What then are the broader implications? First, that carbon emissions worldwide fell, if only marginally, in 1999, attributable wholly to the reduction in Chinese coal use. Second, further comparable reductions in carbon emissions will be possible through the displacement of coal by imported natural gas if current proposals are brought to fruition, benefiting not only China but the whole world.

Another environmentally beneficial development would be to switch from coal to natural gas where possible. Western governments should encourage regional cooperative efforts to promote private-sector investment in natural gas production and infrastructure facilities in the developing countries.[16] Natural gas–fired combined-cycle plants are the most fuel efficient and economic form of electricity generation. But the ability of countries to switch to natural gas is dependent on the construction of expensive natural gas delivery infrastructure in the form of pipelines and liquefied natural gas (LNG) liquefaction and regasification facilities. APEC's natural gas initiative, designed to accelerate private-sector investment in natural gas supply in Asia, as well as the proposed hemispheric natural gas initiative to foster international gas trade in the Western Hemisphere, are steps in the right direction.[17]

Rising electricity demand will be a key indicator of success as the developing countries work to industrialize, and it is here that fuel options must be broadened. As noted, most countries rely on coal, often of poor quality and a major pollutant. Any oil and natural gas to be consumed usually is imported. Hydropower potential is already developed and in use or is environmentally objectionable and other renewables—solar, wind, biomass—fall short in terms of levels of likely future contributions.

The Nuclear Option

Nuclear power can serve as a source of electricity supply free from carbon emissions. Reversing the decline in the generation of nuclear electricity in the developed world, however, poses several policy challenges. One of the major constraints on the greater use of nuclear power in electricity generation comes from higher up-front construction costs compared with natural gas cogeneration facilities. Other constraints include public concerns about the safety of nuclear reactors and safety and environmental problems relative to the disposition of nuclear waste materials that are the by-product of the nuclear energy production process. An additional policy constraint arises from the concern that the nuclear materials and technology used in civil nuclear power programs could be diverted to aid terrorists or countries that are seeking to acquire nuclear weapons outside the bounds of current arms control treaties. Yet, over time, nuclear electric power becomes very competitive with fossil-fuel power generation.

Decisionmakers are thus confronted with a difficult choice: Exercise the nuclear option, through government and private-sector partnership, or accept that pollution will worsen.

For the nuclear option to be acceptable to developing countries, four basic criteria will have to be met. First, the nuclear reactors must be modular and much smaller in generating capacity than their predecessors, perhaps 100 MW or so. Modular reactors will be essential for use in rural areas and responsive to electricity demand growth. Second, nuclear electric power supplied by these fourth genera-

---

16. Bill Richardson, U.S. secretary of energy (remarks at the CSIS conference, "The Geopolitics of Energy into the 21st Century," Washington, D.C., December 9, 1999).

17. Ibid.

tion reactors must be competitive in cost with electricity generated from the burning of fossil fuels. Third, to offset public concern both in the host country and abroad, these reactors must be proliferation resistant. Fourth, as it is with developed countries, the issue of nuclear waste must be satisfactorily resolved.

Western governments should assess the conditions under which nuclear power could make a significant contribution to generating electricity in the developing world. But first that assessment should judge those considerations under which nuclear power could make a contribution to their own supply. In those Western countries where nuclear reactors already are operating, this would require, at a minimum, improving nuclear technology first for use in developed countries, providing technical assistance to establish a culture of nuclear safety at the plants, and strengthening measures to ensure against the diversion of fissile materials. Where nuclear power is absent, cost-competitive and pollution-resistant modular nuclear reactors would offer an attractive choice.

Two Fundamental Considerations

As policymakers attempt to cope with increasing public concerns regarding linkages among energy production and energy consumption and environmental degradation, the following considerations are fundamental.

- Countries should review the extent to which subsidies for domestic energy sectors are inconsistent with their global energy policies. At the same time, governments should review whether existing environmental legislation creates barriers to innovation in environmental technologies and discourages private venture capital commitment to improving the environment.

- OECD governments should sustain and strengthen basic research efforts into new technologies that could make energy use more efficient, improve carbon capture and sequestration, and offer viable alternatives to fossil fuels.

CHAPTER 5

Conclusions

One of the ironies of the turn of the century is that, in an age when the pace of technological change is almost overwhelming, the world is and will remain dependent, for 2000–2020 at least, essentially on the same fossil-fuel sources of energy that prevailed in the twentieth century. Given the nature of these fuels—where they are found and where they are consumed—their production and transportation will continue to be vulnerable to geopolitical upheaval.

The geopolitical risks attendant to energy availability are not expected to abate. Whether war, domestic turmoil, economic recession, or the normal ebb and flow of the political relations between competing nation states, geopolitics and energy will continue to be inextricably intertwined. In sum, there will be recurring geopolitical risk in the twenty-first century, and it will not be possible to isolate energy from this risk.

The challenge for Western policymakers is not to eliminate geopolitical risk but how to manage it. The overarching theme of the SEI's findings is that there are no geopolitical problems affecting energy availability and reliability that cannot be managed, provided decisionmakers in both the public and private sectors first understand the interrelationship between geopolitics and energy and, second, adopt prudent policies that take these interrelationships into account.

Three broad conclusions can be drawn from the SEI analysis.

First, the United States, as the world's only superpower, must accept its special responsibilities for preserving access to worldwide energy supply. This responsibility has implications for military spending needed to develop the appropriate capabilities for power projection and sea-lane protection. It will require an intensification of diplomatic efforts to foster peace and stability in energy-producing regions and to integrate major new importers such as China and India into diversified and reliable energy markets. It also will demand a careful balancing of competing U.S. national interests in policy decisions—such as the implementation of sanctions—that bear importantly on energy. To be sure, the United States should enlist the support and cooperation of its principal allies in sharing the burden of ensuring that energy supplies are adequate and reliable in the future but, regardless, Washington must take the lead.

Second, developing an adequate and reliable energy supply to realize the promise of the globalized twenty-first century will require significant investments now to provide the energy infrastructure needed for the future. Both the size of the investments and the need for their efficient application will thus require heavy reliance on the private sector. Governments and the international financial institutions can usefully support the private-sector effort by providing judicious assistance to risk sharing, where needed, and in encouraging the energy-producing countries to develop an economic climate conducive to investment. In addition, governments have a particular role to play in organizing the protection of the infrastructure critical to energy supply and in the financing and managing of strategic petroleum stocks.

Finally, a special challenge to decisionmakers lies in balancing the objectives of economic growth and environmental protection. Few would dispute that prosperity in the globalized twenty-first century is ineluctably linked to increased energy use. This means, for the foreseeable future, continued reliance on fossil fuels, which in turn holds implications for the environment. Resolving the debate between economic growth and the environment demands a twofold response falling largely to the developed nations, which must, first, develop and bring to market the new technologies that will provide cleaner energy and, second, encourage and facilitate the efforts of the developing countries to choose less environmentally damaging fuels when possible.